STEMD² Research & Development Group

Center on Disability Studies

College of Education

University of Hawai'i at Mānoa

http://stemd2.com/

Copyright ©2020 Center on Disability Studies, University of Hawai'i at Mānoa. All rights reserved. Printed in the United States of America. First published (2017) by the Center on Disability Studies, University of Hawai'i at Mānoa, Honolulu, Hawai'i. This document is based upon work supported by the Department of Education, Native Hawaiian Education Act Program under award #S362A140018. Any opinions, findings, and conclusions or recommendations expressed in this material are those of the authors and do not necessarily reflect the views of the United States Department of Education. For further information about this document and Ne'epapa Ka Hana project, please contact Dr. Kaveh Abhari at abhari@hawaii.edu.

ISBN: 978-0-9983142-4-2

Second release, 2020

Unit 6

1. What are some interesting things that you see about the prices of the "meal packs"?

2. For each meal pack, find out the total cost(s) to feed your family for the week. Fill out the table above. Explain how you found your answer.

The following fishing spots are all around the island, and we have to consider the quality and cost of each location. Your family decides to take your family's truck to transport everything for the trip, and the truck gets *20 miles per gallon*. The gas price is *$3.25 per gallon*.

Here's some information about each fishing spot.

Fishing spot	Description	Distance away	Travel cost
1	Far and hardest to get to, but is full of fish and small wildlife. Secluded and very pretty.	76 miles	$12.35
2	Far, but has a good amount of fish. Clean and not too crowded.	43 miles	$6.99
3	Not too far. The fishing activity is okay. But a little crowded.	22 miles	$3.58
4	Closest spot, but doesn't have a lot of fish. Crowded, a little noisy, and a little dirty.	5 miles	$0.81

Keep in mind that the travel time will affect your fishing and camping time.

3. Calculate how much it will cost to travel to each of the different fishing locations. Fill out the previous table. Explain how you got your answer.

4. Which meal pack(s) and which fishing spot would you choose? Justify your answer and find the cost of your fishing trip.

Activity 6.3

Grade	08
Claim(s)	Claim 1: Concepts and Procedures Students can explain and apply mathematical concepts and carry out mathematical procedures with precision and fluency. Claim 3: Communicating Reasoning Students can clearly and precisely construct viable arguments to support their own reasoning and to critique the reasoning of others.
Assessment Target(s)	3 E: Distinguish correct logic or reasoning from that which is flawed, and—if there is a flaw in the argument—explain what it is. 1 E: Define, evaluate, and compare functions. 1 A (Gr 7): Analyze proportional relationships and use them to solve real-world and mathematical problems. 3 F: Base arguments on concrete referents such as objects, drawings, diagrams, and actions.
Content Domain	Equations and Expressions
Standard(s)	8.F.A.3
DOK	2
Activity Key	*Answers for the table are below. The other answers are open ended.*

Summer fishing trips are always a fun way to spend time with family and friends. As a reward for all your hard work during the school year, you can camp out, fish, and relax before heading back to school. Catching and eating fresh fish makes the experience all the more memorable. As much fun as fishing trips are, there can be a lot of planning that goes into it.

Let's plan a week-long fishing trip for your family of five (including you). You are asked to shop for delicious food and to pick a fun fishing location.

There's a store nearby that sells "meal packs." Three meal packs feed one person for one day.

Meal pack	Description	Price	Total cost for your family
I Stay Broke	Canned beans. Great for saving money!	$2.00	$210.00
Maka'āinana	Spam and rice. The classics!	$2.50	$262.50
Mouth Stay Broke	Kalua pig, salmon, and more. A local favorite!	$3.50	$367.50
Fat Like Wallet	Everything. Never go hungry or unsatisfied. Eat like an ali'i.	$5.00	$525.00

Fish grade	Cost per pound of fish ($)
☆	100
☆☆	110
☆☆☆	121
☆☆☆☆	133.10
☆☆☆☆☆	146.41

After you have finished filling out the table, determine if prices are changing linearly with grade. Please explain why or why not.

Activity 6.2

Grade	08
Claim(s)	Claim 1: Concepts and Procedures Students can explain and apply mathematical concepts and carry out mathematical procedures with precision and fluency. Claim 3: Communicating Reasoning Students can clearly and precisely construct viable arguments to support their own reasoning and to critique the reasoning of others.
Assessment Target(s)	3 E: Distinguish correct logic or reasoning from that which is flawed, and—if there is a flaw in the argument—explain what it is. 1 E: Define, evaluate, and compare functions. 1 A (Gr 7): Analyze proportional relationships and use them to solve real-world and mathematical problems. 3 F: Base arguments on concrete referents such as objects, drawings, diagrams, and actions.
Content Domain	Equations and Expressions
Standard(s)	8.F.A.3
DOK	2
Activity Key	*Answers will vary for the table and an example is included below. The table data cannot be an example of a linear relationship.*

The price of fish changes a lot with "supply and demand." For example, in winter, ahi is normally not biting, so they're hard to catch and the *supply* is low. People want to eat more fish during the holiday season so the *demand* is high. So fresh ahi can get really expensive in the winter months. The "grade" or quality of the fish can also affect its price.

Let's assume that for each higher quality grade of fish, the price increased by 10%. Fill out the table below to create an example of how fish prices could depend on grade, with five star (☆☆☆☆☆) fish being the best, and one star (☆) being the worst. Start by choosing a cost for the one star (☆) fish and work your way up.

The following table shows the amount and weights of catches of oama.

Number of oama caught	Weight of catch (oz)
10	24
15	30
12	23
18	42
22	47
17	34
19	39
25	58
14	26
11	18

Construct a scatter plot of the data in the table on the graph below:

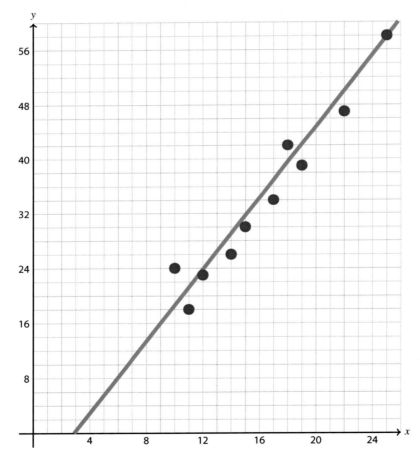

Draw a line of best fit on the graph for the data. Use this line to find how much more oama is needed to increase the weight of a catch by 40 more oz. Explain how you found your answer.

Activity 6.1

Grade	08
Claim(s)	Claim 1: Concepts and Procedures Students can explain and apply mathematical concepts and carry out mathematical procedures with precision and fluency. Claim 2: Problem Solving Students can solve a range of complex well-posed problems in pure and applied mathematics, making productive use of knowledge and problem solving strategies. Claim 3: Communicating Reasoning Students can clearly and precisely construct viable arguments to support their own reasoning and to critique the reasoning of others.
Assessment Target(s)	2 D: Identify important quantities in a practical situation and map their relationships (e.g., using diagrams, two-way tables, graphs, flowcharts, or formulas). 1 J: Investigate patterns of association in bivariate data. 2 A: Apply mathematics to solve well-posed problems arising in everyday life, society, and the workplace. 2 B: Select and use appropriate tools strategically. 2 C: Interpret results in the context of a situation. 3 F: Base arguments on concrete referents such as objects, drawings, diagrams, and actions.
Content Domain	Geometry
Standard(s)	8.SP.A.1, 8.SP.A.2, 8.SP.A.3
DOK	2
Activity Key	*Scatter plot is shown below. About 15 more oama is needed to increase the catch by 40 ounces. We got this answer by utilizing the least squares method to find the line of best fit, $y = 2.62x - 7.6058$ where x is the number of oama and y is the weight. Of course it is perfectly fine to "eyeball" the line of best fit and come up with an answer other than 15, as long as the students justify their work.*

Unit 6: Statistics

Activity 5.9

Grade	08
Claim(s)	Claim 1: Concepts and Procedures Students can explain and apply mathematical concepts and carry out mathematical procedures with precision and fluency. Claim 3: Communicating Reasoning Students can clearly and precisely construct viable arguments to support their own reasoning and to critique the reasoning of others.
Assessment Target(s)	3 B: Construct, autonomously, chains of reasoning that will justify or refute propositions or conjectures. 1 H: Understand and apply the Pythagorean theorem. 3 F: Base arguments on concrete referents such as objects, drawings, diagrams, and actions..
Content Domain	Geometry
Standard(s)	8.G.B.7, 8.G.B.8
Mathematical Practice(s)	2, 3, 6, 7
DOK	4
Activity Key	*Your conjecture is true. The inequality $x < y < z$ is consistent with an obtuse triangle.*

While out on a fishing and camping trip with your friends, a storm began to approach. So you and your friends rush to make a frame for your shelter. Between you and your friends, you have sticks of lengths x, y, and z inches, such that $x < y < z$. You also notice that $x^2 + y^2 < z^2$.

One of your friends say that this is perfect and that you can create a right triangle to build part of your frame for your shelter. You, on the other hand, think that this is not correct and that you will have an obtuse triangle, which won't be good for your shelter since it won't be standing properly. Come up with a reasoning to justify who is right.

Activity 5.8

Grade	08
Claim(s)	Claim 1: Concepts and Procedures Students can explain and apply mathematical concepts and carry out mathematical procedures with precision and fluency. Claim 2: Problem Solving Students can solve a range of complex well-posed problems in pure and applied mathematics, making productive use of knowledge and problem solving strategies. Claim 3: Communicating Reasoning Students can clearly and precisely construct viable arguments to support their own reasoning and to critique the reasoning of others.
Assessment Target(s)	3 F: Base arguments on concrete referents such as objects, drawings, diagrams, and actions. 1 I: Solve real-world and mathematical problems involving volume of cylinders, cones, and spheres. 2 A: Apply mathematics to solve well-posed problems arising in everyday life, society, and the workplace.
Content Domain	Geometry
Standard(s)	8.G.C.9
DOK	3
Activity Key	*Both the sphere and the cylinder have the same volume.*

Floaters are often used in fishing to keep baits and lures off of the ground. Some floaters float better than others and *how much something tends to float* is called *buoyancy*. The more volume an object has, the more buoyancy it also has. Suppose that you have two floaters: one is a sphere with radius of r, and the other is a right cylinder with a radius of r and a height of $\frac{4}{3}r$. Does the sphere have a greater volume, the cylinder have a greater volume, or do they both have the same volume?

Activity 5.7

Grade	08
Claim(s)	Claim 1: Concepts and Procedures Students can explain and apply mathematical concepts and carry out mathematical procedures with precision and fluency. Claim 2: Problem Solving Students can solve a range of complex well-posed problems in pure and applied mathematics, making productive use of knowledge and problem solving strategies.
Assessment Target(s)	2 B: Select and use appropriate tools strategically. 1 I: Solve real-world and mathematical problems involving volume of cylinders, cones, and spheres. 1 B: Work with radicals and integer exponents.
Content Domain	Geometry
Standard(s)	8.G.C.9, 8.EE.A.2
DOK	3
Activity Key	*The radius of the sphere cage should be 12 feet.*

Native Hawaiians have always been excellent at building ponds for farming fish. In modern times, many societies raise fish in large cages or nets in ocean. When fish farming, it is important to consider the volumes of the cages. If we had a cone cage with a height of 27 feet and a base diameter of 32 feet but wanted a sphere cage with the *same volume*, then what should the radius of the sphere be?

Activity 5.6

Grade	08
Claim(s)	Claim 1: Concepts and Procedures Students can explain and apply mathematical concepts and carry out mathematical procedures with precision and fluency. Claim 2: Problem Solving Students can solve a range of complex well-posed problems in pure and applied mathematics, making productive use of knowledge and problem solving strategies.
Assessment Target(s)	2 A: Apply mathematics to solve well-posed problems arising in everyday life, society, and the workplace. 1 I: Solve real-world and mathematical problems involving volume of cylinders, cones and spheres.
Content Domain	Geometry
Standard(s)	8.G.C.9
DOK	1
Activity Key	*Answers will vary*

It is illegal to catch papio that are too small. In order to figure out how long it takes papio to grow to a legal catching size, you decide to raise some baby papio. To raise baby papio, you need to make a *cylindrical* tank with a volume between 100 and 120 cubic feet. Find 3 different possible configurations that will give you your desired tank size.

	Tank #1	Tank #2	Tank #3
Radius (feet)			
Height (feet)			
Volume (feet^3)			

Activity 5.5

Grade	08
Claim(s)	Claim 1: Concepts and Procedures Students can explain and apply mathematical concepts and carry out mathematical procedures with precision and fluency.
Assessment Target(s)	1 B: Work with radicals and integer exponents.
Content Domain	Equations and Expressions
Standard(s)	8.EE.A.2
DOK	2
Activity Key	A. 256 in^3 and 361 in^3 B. 1 ft^3 and 64 in^3 C. 125 in^3 D. 96 in^3, 200 in^3, and 333 in^3

You are looking online to buy a new cooler for your fishing trip. The website you are looking at has the following description of their coolers:

"These new *cube* coolers are perfect for fishing trips!. Get your coolers with the following volumes: 1 ft^3, 64 in^3, 96 in^3, 125 in^3, 200 in^3, 256 in^3, 333 in^3, 361 in^3!"

You want a cooler that has *integer dimensions*. Sort the coolers out in the following table to help you decide which cooler to get.

A. Perfect square volumes (not perfect cubes)	B. Both perfect square and perfect cube volumes	C. Perfect cube volumes (not perfect squares)	D. Neither perfect cube nor perfect square volumes

Activity 5.4

Grade	08
Claim(s)	Claim 1: Concepts and Procedures Students can explain and apply mathematical concepts and carry out mathematical procedures with precision and fluency.
Assessment Target(s)	1 B: Work with radicals and integer exponents.
Content Domain	Equations and Expressions
Standard(s)	8.EE.A.2
DOK	1
Activity Key	1. *The box should be 8 in $\times$ 8 in $\times$ 1 in.* 2. *The box should be 10 in $\times$ 10 in $\times$ 10 in.*

1. Fishing gear takes up a lot of room and organization really helps with getting your fishing set up ready. One common item among all fishermen is a tackle box. It is a box used to organize fishing gear such as hooks, swivels, lead, etc. Say you wanted to make your own tackle box which has a square bottom, a height of 1 inch, and a volume of 64 cubic inches. What should the dimensions of your final tackle box be?

2. Say you wanted a tackle box for your lures, which are quite big, and you wanted your box to be a cube with a volume of 1,000 cubic inches. What should the dimensions of the lure tackle box be?

Activity 5.3

Grade	08
Claim(s)	Claim 1: Concepts and Procedures Students can explain and apply mathematical concepts and carry out mathematical procedures with precision and fluency.
Assessment Target(s)	1 H: Understand and apply the Pythagorean theorem.
Content Domain	Geometry
Standard(s)	8.G.B.7, 8.G.B.8
DOK	1
Activity Key	Answers will vary.

When constructing a speargun, it is easiest to start with a block of wood that is a rectangular prism, like a long shoe box. Every angle of a rectangular prism is perpendicular, and you can check your angles with a special tool called a *square*. If you do not have a square, then you can make one with Pythagorean triples. We just need to get 3 pieces of wood each cut to a length of a, b, and c, such that $a^2 + b^2 = c^2$.

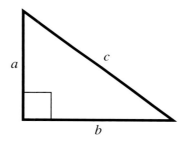

Find *two* sets of numbers (measurements) that our a, b, and c sticks might have to make our square tool.

Activity 5.2

Grade	08
Claim(s)	Claim 1: Concepts and Procedures Students can explain and apply mathematical concepts and carry out mathematical procedures with precision and fluency.
Assessment Target(s)	1 G: Understand congruence and similarity using physical models, transparencies, or geometry software. 1 D: Analyze and solve linear equations and pairs of simultaneous linear equations.
Content Domain	Geometry
Standard(s)	8.G.A.5, 8.EE.C.7
DOK	1
Activity Key	$x = 70°, y = 60°,$ and $z = 60°$

Ancient Hawaiians didn't use fishing reels—not even when catching *ulua* (a large predatory reef fish). Instead, Hawaiians used a traditional method was called *kau lāʻau* "hang stick". Check out this YouTube video on Hawaiian ulua fishing: http://bit.ly/2pTqTs2.

Kau lāʻau involves using a long stick with a rope at the end, which hangs a bait directly below it. Below is an image of a modern pole, represented by the line segment AB and its fishing line represented by BD. The pole used in the kau lāʻau is represented by AC and it's line CD. Find the angles x, y, and z.

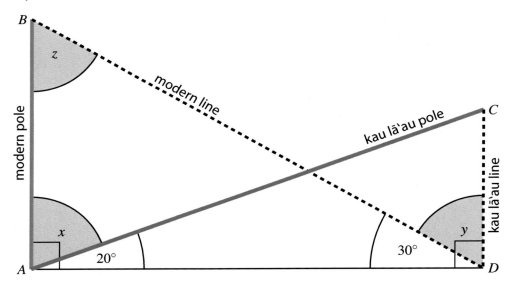

Activity 5.1

Grade	08
Claim(s)	Claim 1: Concepts and Procedures Students can explain and apply mathematical concepts and carry out mathematical procedures with precision and fluency.
Assessment Target(s)	1 H: Understand and apply the Pythagorean theorem.
Content Domain	Geometry
Standard(s)	8.G.B.8
DOK	2
Activity Key	*17 feet*

You want to lay a net in a straight line from where you are on the water's edge to a point that is 15 feet straight out and 8 feet to that position's right. How long a net do you need?

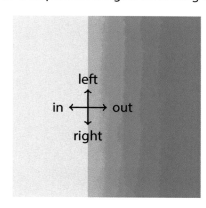

Unit 5: Measurement Geometry

"If you reflect a polygon over the y-axis, then you just gotta keep the same y-coordinates, but take the opposite of the x-coordinates."

Is your friend's conjecture true?

If it is true, explain why the x and y coordinates become opposite/stay the same. You may also graph a few examples to make your explanation clearer. *If it is false*, provide an example.

Activity 4.5

Grade	08
Claim(s)	Claim 1: Concepts and Procedures Students can explain and apply mathematical concepts and carry out mathematical procedures with precision and fluency. Claim 2: Problem Solving Students can solve a range of complex well-posed problems in pure and applied mathematics, making productive use of knowledge and problem solving strategies. Claim 3: Communicating Reasoning Students can clearly and precisely construct viable arguments to support their own reasoning and to critique the reasoning of others.
Assessment Target(s)	3 B: Construct, autonomously, chains of reasoning that will justify or refute propositions or conjectures. 1 G: Understand congruence and similarity using physical models, transparencies, or geometry software. 2 B: Select and use appropriate tools strategically. 3 F: Base arguments on concrete referents such as objects, drawings, diagrams, and actions.
Content Domain	Geometry
Standard(s)	8.G.A.3
DOK	4
Activity Key	*This is true.*

While designing your new canoe, you and your friend agree that the canoe should be symmetric when looking at it from the front.

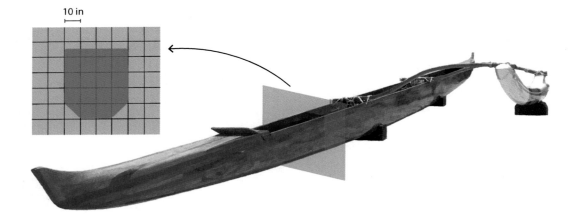

The two of you are sketching the design out on graph paper and discussing about the best way to make sure that it is symmetric. Then your friend says:

1. Create the second month's net placement by drawing a similar triangle to triangle ABC. Label this new triangle XYZ.

2. Describe how your second month's net layout could be obtained from triangle ABC with mathematical transformations.

3. Describe the mathematical transformation to obtain triangle XYZ from triangle FED.

4. Are the blue and orange triangles congruent? Explain your reasoning.

Unit 4

Activity 4.4

Grade	08
Claim(s)	Claim 1: Concepts and Procedures Students can explain and apply mathematical concepts and carry out mathematical procedures with precision and fluency. Claim 2: Problem Solving Students can solve a range of complex well-posed problems in pure and applied mathematics, making productive use of knowledge and problem solving strategies.
Assessment Target(s)	2 C: Interpret results in the context of a situation. 2 A: Apply mathematics to solve well-posed problems arising in everyday life, society, and the workplace. 1 F: Use functions to model relationships between quantities.
Content Domain	Geometry
Standard(s)	8.G.A.1, 8.G.A.1.a, 8.G.A.1.b, 8.G.A.1.c, 8.G.A.2, 8.G.A.3, 8.G.A.4
DOK	3
Activity Key	*1-3. Answers will vary.* *4. They are similar but not congruent since at least one dilation is needed to transform one triangle to the other.*

You noticed over several months that your fishing grounds have been changing due to the current and waves. In the first month, you placed your net as the blue triangle (ABC) below to surrounded a school of fish. In the second month, you had to move your net by a mathematical transformation. Finally, during the third month, you laid your net with the orange triangle (DEF) representing your net placement. Your net placements were similar triangles in all three months.

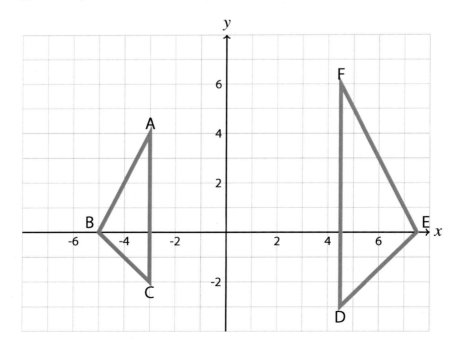

Activity 4.3

Grade	08
Claim(s)	Claim 1: Concepts and Procedures Students can explain and apply mathematical concepts and carry out mathematical procedures with precision and fluency.
Assessment Target(s)	1 G: Understand congruence and similarity using physical models, transparencies, or geometry software.
Content Domain	Geometry
Standard(s)	8.G.A.1.a, 8.G.A.1.b, 8.G.A.1.c, 8.G.A.2, 8.G.A.4
DOK	1
Activity Key	*New points should be at: (-3, 0), (-3, 4), (0, 0), (1, 4).*

Rubbish piles floating in the ocean often attract small fish, which seek shelter under the rubbish. These small fish also attract larger predatory fish, which makes the rubbish piles good fishing locations. The shape on the graph below represents a rubbish pile. In a month, it will drift 4 miles south, and 5 miles west from its current location. Find the rubbish pile's new location and draw the rubbish pile on the graph.

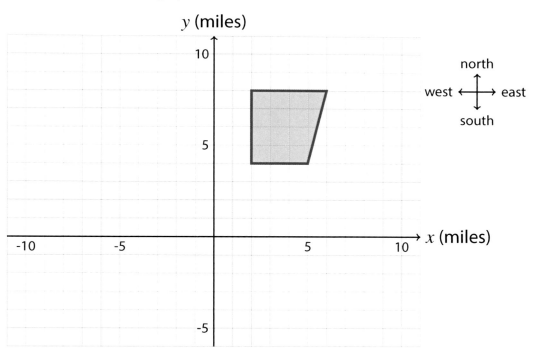

Unit 4

Activity 4.2

Grade	08
Claim(s)	Claim 1: Concepts and Procedures Students can explain and apply mathematical concepts and carry out mathematical procedures with precision and fluency.
Assessment Target(s)	1 G: Understand congruence and similarity using physical models, transparencies, or geometry software.
Content Domain	Geometry
Standard(s)	8.G.A.2
DOK	3
Activity Key	AB = Y, AC = Z, BC = X

Lay nets are placed in triangles surrounding two different fishing holes in a bay.

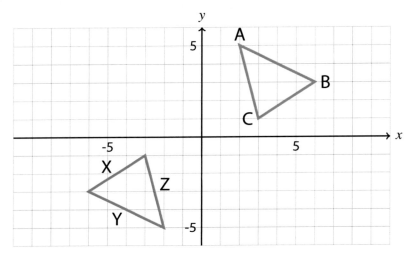

Check the appropriate boxes in the table to show which sides of the triangles have equal length.

	X	Y	Z
AB			
AC			

Activity 4.1

Grade	08
Claim(s)	Claim 1: Concepts and Procedures Students can explain and apply mathematical concepts and carry out mathematical procedures with precision and fluency.
Assessment Target(s)	1 G: Understand congruence and similarity using physical models, transparencies, or geometry software.
Content Domain	Geometry
Standard(s)	8.G.A.3, 8.G.A.4
DOK	1
Activity Key	Answers will vary, but since the point B is at (6, 3), when we dilate the shape about the origin, B can move to (6n, 3n) where n is any real positive number.

Lay nets are a type of rectangular net with weights along one edge and floats on the opposite edge. When you lay it in the water, it stands up like a fence. You usually set up your lay net in a triangle like in the picture below.

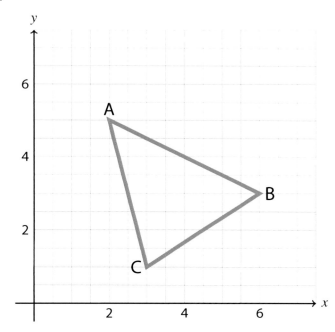

Your uncle gave you a larger lay net. Now that you have more netting material, you decide to take a look at your original setup and dilate it about the origin.

What could be a new possible location of point B in the bay? Please explain your reasoning. Is this the only possible position? Explain.

Unit 4: Transformational Geometry

Unit 3

Activity 3.5

Grade	08
Claim(s)	Claim 1: Concepts and Procedures Students can explain and apply mathematical concepts and carry out mathematical procedures with precision and fluency. Claim 2: Problem Solving Students can solve a range of complex well-posed problems in pure and applied mathematics, making productive use of knowledge and problem solving strategies.
Assessment Target(s)	2 A: Apply mathematics to solve well-posed problems arising in everyday life, society, and the workplace. 2 B: Select and use appropriate tools strategically. 1 F: Use functions to model relationships between quantities..
Content Domain	Equations and Expressions
Standard(s)	8.EE.B.5, 8.EE.B.6
DOK	2
Activity Key	*Answers will vary. An example is given below.*

Your family is on a fishing trip, and you decide to catch something for dinner. Your favorite fish to eat is papio, which live in a channel that runs diagonally to the beach where you're standing. There's a particularly nice fishing spot in the channel, shown on the graph at point (4, -2). You want to cast your line from the beach straight to the fishing spot. You also want the slope of your fishing line's equation to be *greater than -1*. Graph two possible fishing lines.

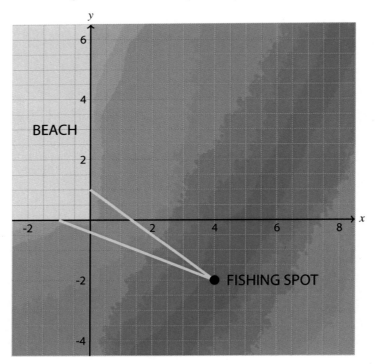

Activity 3.4

Grade	08
Claim(s)	Claim 1: Concepts and Procedures Students can explain and apply mathematical concepts and carry out mathematical procedures with precision and fluency. Claim 2: Problem Solving Students can solve a range of complex well-posed problems in pure and applied mathematics, making productive use of knowledge and problem solving strategies.
Assessment Target(s)	2 C: Interpret results in the context of a situation. 2 A: Apply mathematics to solve well-posed problems arising in everyday life, society, and the workplace. 1 F: Use functions to model relationships between quantities.
Content Domain	Functions
Standard(s)	8.EE.C.7, 8.EE.C.8, 8.EE.C.8.b, 8.EE.C.8.c, 8.F.B.4
DOK	3
Activity Key	*Cost per hook:* 0.02 *Shipping fee:* 3.00 *Cost for n hooks:* $\$0.02 \times n + \$3.00$

Suppose that you want to buy fishing hooks online, and the shipping cost is always the same no matter how many hooks you buy at a time.

Last month, you bought 500 hooks for $13.00 (including shipping). This month, you bought 5000 hooks for $103.00 (including shipping).

What is the cost for each hook, not including the shipping charge?

What is the cost of the shipping?

What is the cost for n, number of hooks?

Activity 3.3

Grade	08
Claim(s)	Claim 1: Concepts and Procedures Students can explain and apply mathematical concepts and carry out mathematical procedures with precision and fluency.
Assessment Target(s)	1 D: Analyze and solve linear equations and pairs of simultaneous linear equations.
Content Domain	Equations and Expressions
Standard(s)	8.EE.C.8.a
DOK	3
Activity Key	Answers will vary. An example is given.

A modern GPS (Global Positioning System) can accurately pinpoint your location and track your progress throughout a day. Your GPS data shows that on your last two fishing trips, you traveled in straight lines and found a great fishing spot on both days at the point (-3, 2). To get to that point quickly, you decide to travel in a straight line from the shore (represented by the x-axis). Find *two* possible paths to get to your fishing spot (write the equations). Graph your paths and the point where you want to end up.

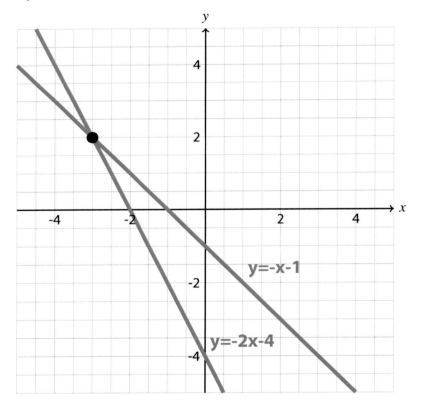

Did you just form a system of linear equations? How do you know?

Below is the bay with a Cartesian plane. Check your answers by graphing the lines on the plane.

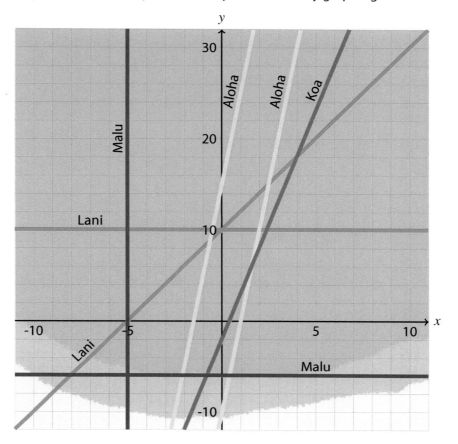

Unit 3

Activity 3.2

Grade	08
Claim(s)	Claim 1: Concepts and Procedures Students can explain and apply mathematical concepts and carry out mathematical procedures with precision and fluency.
Assessment Target(s)	1 D: Analyze and solve linear equations and pairs of simultaneous linear equations.
Content Domain	Equations and Expressions
Standard(s)	8.EE.C.7, 8.EE.C.7.a, 8.EE.C.7.b, 8.EE.C.8, 8.EE.C.8.a, 8.EE.C.8.b, 8.EE.C.8.c
DOK	2
Activity Key	See table and graph below

Among the fishing community, it is common to keep your favorite fishing spots a secret. This prevents overfishing at these spots. A couple of your friends have decided to shared their favorite spots in a bay by giving two equations each—their fishing spot is supposed to be at the intersection of the two equations. Find the intersections of the following equations (if you can!), and see if they were honest or lying in telling you their secret spot.

Friend	Equation 1	Equation 2	Solution	Lying or honest?
Lani	$y = 2x + 10$	$y = 10$	(0, 10)	Honest
Aloha	$y = 10x + 15$	$y = 10x - 10$	no solution	Lying
Koa	$y = 5x - 2$	$y = \dfrac{15x - 6}{3}$	infinitely many solutions	Lying
Malu	$y = -6$	$x = -5$	(-5, -6)	Honest

Activity 3.1

Grade	08
Claim(s)	Claim 1: Concepts and Procedures Students can explain and apply mathematical concepts and carry out mathematical procedures with precision and fluency.
Assessment Target(s)	1 D: Analyze and solve linear equations and pairs of simultaneous linear equations.
Content Domain	Equations and Expressions
Standard(s)	8.EE.C.7, 8.EE.C.7.b
DOK	3
Activity Key	*Arm length is 24 inches*

When bottom fishing, some fishermen use their arm length to estimate the lengths of their fishing lines. A normal bottom fishing set-up consists of taking your arm length and adding 6 inches to that. You then double this length to make 60 inches. Write out an algebraic equation to describe this, and use that equation to find how long your arm length is.

Unit 3: Solving Equations and Systems of Equations

Lilo saved up some money too, but showed you her savings in a table:

Week	Total amount saved
1	$4.50
2	$9.00
3	$13.50
4	$18.00

You also saved some money, but you saved the same amount of money every week for 4 weeks. You modeled your savings with the equation: $s = 5.5w$, where s is how much you saved and w is the number of weeks.

Among the three of you, who was able to save the most amount of money each week, and who saved the least amount per week? Please explain how you got your answer.

Unit 2

Activity 2.7

Grade	08
Claim(s)	Claim 1: Concepts and Procedures Students can explain and apply mathematical concepts and carry out mathematical procedures with precision and fluency. Claim 2: Problem Solving Students can solve a range of complex well-posed problems in pure and applied mathematics, making productive use of knowledge and problem solving strategies.
Assessment Target(s)	2 C: Interpret results in the context of a situation. 1 C: Understand the connections between proportional relationships, lines, and linear equations.
Content Domain	Equations and Expressions
Standard(s)	8.F.A.2, 8.F.B.4
DOK	3
Activity Key	*You saved the most, then Kainoa, and lastly Lilo.*

You and two of your friends are saving up for some new fishing gear. Kainoa saved up his money and showed you how much he saved on a graph.

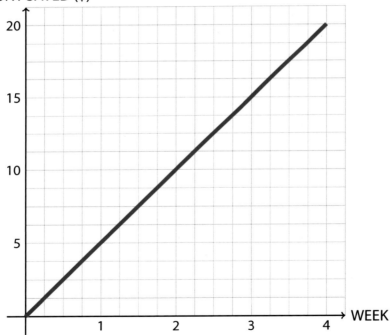

Activity 2.6

Grade	08
Claim(s)	Claim 1: Concepts and Procedures Students can explain and apply mathematical concepts and carry out mathematical procedures with precision and fluency. Claim 2: Problem Solving Students can solve a range of complex well-posed problems in pure and applied mathematics, making productive use of knowledge and problem solving strategies.
Assessment Target(s)	2 A: Apply mathematics to solve well-posed problems arising in everyday life, society, and the workplace. 2 B: Select and use appropriate tools strategically. 1 F: Use functions to model relationships between quantities..
Content Domain	Functions
Standard(s)	8.F.B.5
DOK	2
Activity Key	*Answers will vary. An example solution is included below.*

Oama season has come again this summer, and here is how the day went fishing for you:

- For the first hour, the fish were biting fast, then it slowed down, You caught 10 oama.
- Over the next two hours, the fish were biting at a steady rate until you had 25 oama total in your cooler.
- Unfortunately, for the next hour, the fish weren't biting at all.
- In the hour after that, the fish started biting slowly. Luckily, the fishing picked up, and you now 40 in the cooler.
- Over your final hour fishing, you steadily caught your last 10 oama, reaching your bag limit for the day.

On the Cartesian plane below, graph how this day of fishing might look in terms of the number of fish caught over time.

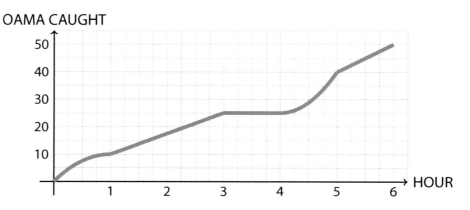

Unit 2

Activity 2.5

Grade	08
Claim(s)	Claim 1: Concepts and Procedures Students can explain and apply mathematical concepts and carry out mathematical procedures with precision and fluency. Claim 2: Problem Solving Students can solve a range of complex well-posed problems in pure and applied mathematics, making productive use of knowledge and problem solving strategies.
Assessment Target(s)	2 A: Apply mathematics to solve well-posed problems arising in everyday life, society, and the workplace. (DOK 2, 3) 1 I: Solve real-world and mathematical problems involving volume of cylinders, cones and spheres.
Content Domain	Geometry
Standard(s)	8.G.C.9
DOK	2
Activity Key	*5 minutes*

A *live well* is a container with a water pump used for keeping fish alive, typically fish that will be used as bait. Suppose that your live well is cylindrical with a height of 10 inches and a radius of 4 inches. Your water pump can bring in new salt water at a rate of 100 cubic inches per minute. At this rate, how long will it take you to completely fill your live well without it overflowing? Round your answer to the nearest minute.

Activity 2.4

Grade	08
Claim(s)	Claim 1: Concepts and Procedures Students can explain and apply mathematical concepts and carry out mathematical procedures with precision and fluency.
Assessment Target(s)	1 C: Understand the connections between proportional relationships, lines, and linear equations.
Content Domain	Equations and Expressions
Standard(s)	8.F.A.1, 8.F.A.3, 8.F.B.4
DOK	1
Activity Key	Red: $y = 2x - 1$ Blue: $y = -\frac{1}{3}x + 2$ Yes these are both examples of linear functions.

A modern GPS (Global Positioning System) can accurately pinpoint your location and track your progress throughout a day. Below is the GPS data from your last two fishing trips.

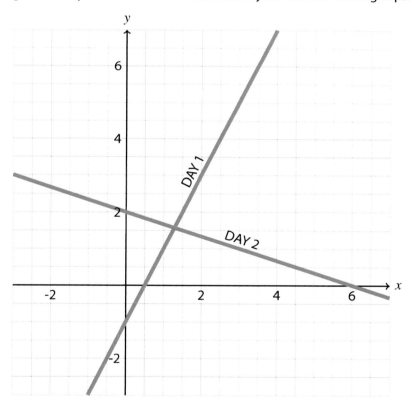

Find the equations to best describe your path for the Day 1 (the orange line) and Day 2 (the blue line) to let your friends know where the fish was biting. Are these equations examples of linear functions?

Unit 2

Activity 2.3

Grade	08
Claim(s)	Claim 1: Concepts and Procedures Students can explain and apply mathematical concepts and carry out mathematical procedures with precision and fluency.
Assessment Target(s)	1 E: Define, evaluate, and compare functions.
Content Domain	Functions
Standard(s)	8.F.A.1
DOK	1
Activity Key	*Any coordinate along the line x=2.*

Finding fish can be hard at times, but there are places where fish tend to hang out. One of these places is along ledges and steep drop offs of the reef. Below is a bird's eye view of a reef with a line showing its drop off. The circle at (2, 4) is a good place to catch fish. What are three other coordinates that are good places to find fish?

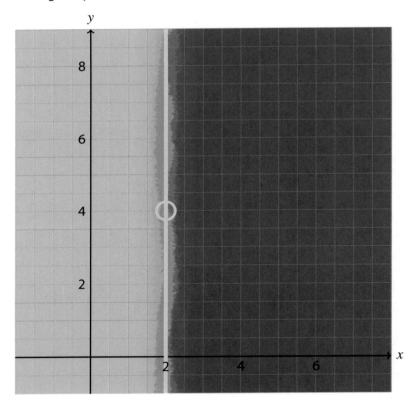

Each of the following graphs represent the amount of oama you need catch for the day (y) vs time (x). Match each graph with a story from above (1-4).

Graph A

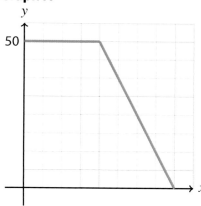

Graph B

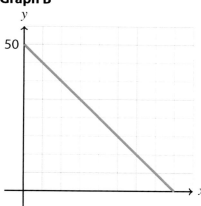

Graph C

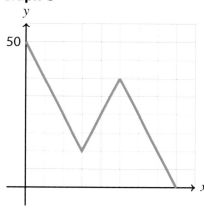

Graph D

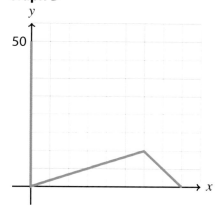

Unit 2

Activity 2.2

Grade	08
Claim(s)	Claim 1: Concepts and Procedures Students can explain and apply mathematical concepts and carry out mathematical procedures with precision and fluency. Claim 2: Problem Solving Students can solve a range of complex well-posed problems in pure and applied mathematics, making productive use of knowledge and problem solving strategies
Assessment Target(s)	1 F: Use functions to model relationships between quantities. 2 C: Interpret results in the context of a situation. 2 A: Apply mathematics to solve well-posed problems arising in everyday life, society, and the workplace. 2 D: Identify important quantities in a practical situation and map their relationships (e.g., using diagrams, two-way tables, graphs, flowcharts, or formulas).
Content Domain	Functions
Standard(s)	8.F.B.5
DOK	2
Activity Key	1. Graph A 2. Graph D 3. Graph B 4. Graph C

There's a graduation party coming up at the end of the summer, and you want to provide fish for your friends and family who attend. So you decide to go fishing for *oama* (juvenile goatfish) the week before the party. You plan to catch the bag limit of 50 oama per day to get to your goal of 200 oama for the party.

Here's a recap of how each fishing day went:

1. The oama weren't biting for half the day, but then the pace of your catch picked up nicely and consistently.

2. As soon as you got to your fishing spot, a nice fishermen gave you 50 oama because he had caught too many. Instead of taking it all home, you decided to use some for bait and fish for fun. At the end of the day, you got back to fishing for oama to replace the ones that you've used.

3. The oama was biting consistently throughout the day and you caught your limit for the day.

4. The oama were biting well and you caught 30 oama quickly, but then you gave some away to another fishermen. Later, you finished catching your limit.

Activity 2.1

Grade	08
Claim(s)	Claim 1: Concepts and Procedures Students can explain and apply mathematical concepts and carry out mathematical procedures with precision and fluency.
Assessment Target(s)	1 E: Define, evaluate, and compare functions.
Content Domain	Functions
Standard(s)	8.F.A.1
DOK	4
Activity Key	*Answers will vary. An example is given below.*

It is usually much safer to go fishing with your friends than by yourself. A lot of dangerous things can happen while out on the ocean, so it's good to have some friends around who can help. For example, a common rule for spearfishing is to have one diver underwater while their partner stays on the surface, waiting to help if needed. Fill out the two tables below. Both tables should show that more lives are saved when there are more people present. The first table should show a function between *people present* and *lives saved*, but the second table should not be a function.

Function

People present (number)	Lives saved (number)
1	0
2	1
3	2
4	3
5	4
7	5
9	5

VS.

Not a Function

People present (number)	Lives saved (number)
1	0
1	1
3	2
4	3
5	4
7	5
9	5

Unit 2: Proportional and Nonproportional Relationships and Functions

Unit 1

Activity 1.4

Grade	08
Claim(s)	Claim 1: Concepts and Procedures Students can explain and apply mathematical concepts and carry out mathematical procedures with precision and fluency. Claim 2: Problem Solving Students can solve a range of complex well-posed problems in pure and applied mathematics, making productive use of knowledge and problem solving strategies.
Assessment Target(s)	2 A: Apply mathematics to solve well-posed problems arising in everyday life, society, and the workplace. 2 B: Select and use appropriate tools strategically. 1 F: Use functions to model relationships between quantities..
Content Domain	Equations and Expressions
Standard(s)	8.EE.A.4
DOK	2
Activity Key	See below for completed table.

The price for fresh poke on the island varies with the supply of fish available. Some of these fish are bought fresh from local fishermen. Unfortunately, today the market's scale was broken and can only weigh things in grams. The worker informs you that today's market price for ahi is $17.64 per kg ($1 \times 10^3$g = 1kg).

The following table shows a great day's catch (4 ahi) and how much you're offered to sell your fish to the market. Wanting to become a more experienced fisherman who sells fish to the market, you would like to calculate the ahi's weight for future reference. Does each fish weigh between 4.219×10^4 and 53.21×10^3 grams?

Fish	Price of fish ($)	Weight of fish (grams)	Between 4.219×10^4 and 53.21×10^3? (Write Yes or No)
#1	800.00	45351.47	Yes
#2	920.20	52154.19	Yes
#3	760.00	43083.90	Yes
#4	640.00	36281.17	No

Activity 1.3

Grade	08
Claim(s)	Claim 1: Concepts and Procedures Students can explain and apply mathematical concepts and carry out mathematical procedures with precision and fluency. Claim 2: Problem Solving Students can solve a range of complex well-posed problems in pure and applied mathematics, making productive use of knowledge and problem solving strategies.
Assessment Target(s)	2 A: Apply mathematics to solve well-posed problems arising in everyday life, society, and the workplace. 1 B: Work with radicals and integer exponents. 1 A: Know that there are numbers that are not rational, and approximate them by rational numbers..
Content Domain	Expressions and Equations The Number System
Standard(s)	8.EE.A.1, 8.EE.A.2, 8.EE.A.3, 8.EE.A.4, 8.NS.A.2
DOK	2
Activity Key	$d = 5.9 \times 10^7$ millimeters

The area of the island of Oahu can be estimated by looking at its greatest dimensions, which is about 7×10^7 millimeters long and about 5×10^7 millimeters wide. If you wanted to make a square net with *enough area to cover the entire island*, how long should each side be? So if d^2 is the area of the island of Oahu, solve the following to determine the length d. Please round your answer to the nearest million millimeters, then write it in scientific notation.
Hint: Area of net = $d^2 = (7 \times 10^7)(5 \times 10^7)$.

Unit 1

Activity 1.2

Grade	08
Claim(s)	Claim 1: Concepts and Procedures Students can explain and apply mathematical concepts and carry out mathematical procedures with precision and fluency.
Assessment Target(s)	1 B: Work with radicals and integer exponents.
Content Domain	Equations and Expressions
Standard(s)	8.EE.A.1
DOK	3
Activity Key	*See below for the completed table. Robert has the highest ratio.*

You and your friends have kept track of all the hours each of you went fishing for the past 5 years (Y), and how many fish each of you caught (X). Find out who has the highest ratio of *number of fish caught* to *number of hours spent fishing* by filling out the table and comparing their ratios. Keep in mind there are different types of fishing, some people scoop nehu (small bait fish) and count that too.

	Number of fish caught (X)	Number of hours spent fishing (Y)	Ratio ($\frac{X}{Y}$)
You	3^4	3^5	$3^{-1} = 0.\overline{3}$
Robert	2^2	2^{-2}	$2^4 = 16$
Sara	2	$(\frac{1}{2})^2$	$2^3 = 8$
Mike	$(-5)^4$	$(-5)^6$	$(-5)^{-2} = 0.04$

Who had the highest ratio?

Activity 1.1

Grade	08
Claim(s)	Claim 1: Concepts and Procedures Students can explain and apply mathematical concepts and carry out mathematical procedures with precision and fluency.
Assessment Target(s)	1 B: Work with radicals and integer exponents
Content Domain	Expressions and Equations
Standard(s)	8.EE.A.3, 8.EE.A.4
DOK	2
Activity Key	*Answer should be between 3.5-3.7 times higher.*

Mauna Kea is about 1.38×10^4 feet tall. Yellowfin tuna, or better known as ahi in the islands, have been recorded to reach depths of 3.8×10^3 feet in the ocean. How much higher is Mauna Kea than the recorded depth an ahi can reach?

Unit 1: The Number System

COMMON CORE STATE STANDARD	FISHING ACTIVITY
8.SP.A.3 Use the equation of a linear model to solve problems in the context of bivariate measurement data, interpreting the slope and intercept. For example, in a linear model for a biology experiment, interpret a slope of 1.5 cm/hr as meaning that an additional hour of sunlight each day is associated with an additional 1.5 cm in mature plant height.	6.1
8.SP.A.4 Understand that patterns of association can also be seen in bivariate categorical data by displaying frequencies and relative frequencies in a two-way table. Construct and interpret a two-way table summarizing data on two categorical variables collected from the same subjects. Use relative frequencies calculated for rows or columns to describe possible association between the two variables. For example, collect data from students in your class on whether or not they have a curfew on school nights and whether or not they have assigned chores at home. Is there evidence that those who have a curfew also tend to have chores?	

COMMON CORE STATE STANDARD	**FISHING ACTIVITY**
8.G.A.5 Use informal arguments to establish facts about the angle sum and exterior angle of triangles, about the angles created when parallel lines are cut by a transversal, and the angle-angle criterion for similarity of triangles. For example, arrange three copies of the same triangle so that the sum of the three angles appears to form a line, and give an argument in terms of transversals why this is so.	5.2
Understand and apply the Pythagorean Theorem.	
8.G.B.6 Explain a proof of the Pythagorean Theorem and its converse.	
8.G.B.7 Apply the Pythagorean Theorem to determine unknown side lengths in right triangles in real-world and mathematical problems in two and three dimensions.	5.3, 5.9
8.G.B.8 Apply the Pythagorean Theorem to find the distance between two points in a coordinate system.	5.1, 5.3, 5.9
Solve real-world and mathematical problems involving volume of cylinders, cones, and spheres.	
8.G.C.9 Know the formulas for the volumes of cones, cylinders, and spheres and use them to solve real-world and mathematical problems.	5.6, 2.5, 5.7, 5.8
Statistic and Probability	
Investigate patterns of association in bivariate data.	
8.SP.A.1 Construct and interpret scatter plots for bivariate measurement data to investigate patterns of association between two quantities. Describe patterns such as clustering, outliers, positive or negative association, linear association, and nonlinear association.	6.1
8.SP.A.2 Know that straight lines are widely used to model relationships between two quantitative variables. For scatter plots that suggest a linear association, informally fit a straight line, and informally assess the model fit by judging the closeness of the data points to the line.	6.1

Continued on next page

COMMON CORE STATE STANDARD	FISHING ACTIVITY
Use functions to model relationships between quantities.	
8.F.B.4 Construct a function to model a linear relationship between two quantities. Determine the rate of change and initial value of the function from a description of a relationship or from two (x, y) values, including reading these from a table or from a graph. Interpret the rate of change and initial value of a linear function in terms of the situation it models, and in terms of its graph or a table of values.	2.4, 3.4, 2.7
8.F.B.5 Describe qualitatively the functional relationship between two quantities by analyzing a graph (e.g., where the function is increasing or decreasing, linear or nonlinear). Sketch a graph that exhibits the qualitative features of a function that has been described verbally.	2.2, 2.6
Geometry	
Understand congruence and similarity using physical models, transparencies, or geometry software.	
8.G.A.1 Verify experimentally the properties of rotations, reflections, and translations:	4.4
8.G.A.1.a Lines are taken to lines, and line segments to line segments of the same length.	4.3, 4.4
8.G.A.1.b Angles are taken to angles of the same measure.	4.3, 4.4
8.G.A.1.c Parallel lines are taken to parallel lines.	4.3, 4.4
8.G.A.2 Understand that a two-dimensional figure is congruent to another if the second can be obtained from the first by a sequence of rotations, reflections, and translations; given two congruent figures, describe a sequence that exhibits the congruence between them.	4.2, 4.3, 4.4
8.G.A.3 Describe the effect of dilations, translations, rotations, and reflections on two-dimensional figures using coordinates.	4.1, 4.4, 4.5
8.G.A.4 Understand that a two-dimensional figure is similar to another if the second can be obtained from the first by a sequence of rotations, reflections, translations, and dilations; given two similar two-dimensional figures, describe a sequence that exhibits the similarity between them.	4.1, 4.3, 4.4

Continued on next page

COMMON CORE STATE STANDARD	FISHING ACTIVITY
8.EE.C.7.b Solve linear equations with rational number coefficients, including equations whose solutions require expanding expressions using the distributive property and collecting like terms.	3.1, 3.2
8.EE.C.8 Analyze and solve pairs of simultaneous linear equations.	3.2, 3.4
8.EE.C.8.a Understand that solutions to a system of two linear equations in two variables correspond to points of intersection of their graphs, because points of intersection satisfy both equations simultaneously.	3.2, 3.3
8.EE.C.8.b Solve systems of two linear equations in two variables algebraically, and estimate solutions by graphing the equations. Solve simple cases by inspection. For example, $3x + 2y = 5$ and $3x + 2y = 6$ have no solution because $3x + 2y$ cannot simultaneously be 5 and 6.	3.2, 3.4
8.EE.C.8.c Solve real-world and mathematical problems leading to two linear equations in two variables. For example, given coordinates for two pairs of points, determine whether the line through the first pair of points intersects the line through the second pair.	3.2, 3.4

Functions

Define, evaluate, and compare functions.

8.F.A.1 Understand that a function is a rule that assigns to each input exactly one output. The graph of a function is the set of ordered pairs consisting of an input and the corresponding output.	2.1, 2.3, 2.4
8.F.A.2 Compare properties of two functions each represented in a different way (algebraically, graphically, numerically in tables, or by verbal descriptions).For example, given a linear function represented by a table of values and a linear function represented by an algebraic expression, determine which function has the greater rate of change.	2.7
8.F.A.3 Interpret the equation $y = mx + b$ as defining a linear function, whose graph is a straight line; give examples of functions that are not linear. For example, the function $A = s^2$ giving the area of a square as a function of its side length is not linear because its graph contains the points (1,1), (2,4) and (3,9), which are not on a straight line.	2.4, 6.3

Continued on next page

COMMON CORE STATE STANDARD	FISHING ACTIVITY
8.EE.A.3 Use numbers expressed in the form of a single digit times an integer power of 10 to estimate very large or very small quantities, and to express how many times as much one is than the other. For example, estimate the population of the United States as 3 times 10^8 and the population of the world as 7 times 10^9, and determine that the world population is more than 20 times larger.	1.1, 1.3
8.EE.A.4 Perform operations with numbers expressed in scientific notation, including problems where both decimal and scientific notation are used. Use scientific notation and choose units of appropriate size for measurements of very large or very small quantities (e.g., use millimeters per year for seafloor spreading). Interpret scientific notation that has been generated by technology	1.1, 1.3, 1.4
Understand the connections between proportional relationships, lines, and linear equations.	
8.EE.B.5 Graph proportional relationships, interpreting the unit rate as the slope of the graph. Compare two different proportional relationships represented in different ways. For example, compare a distance-time graph to a distance-time equation to determine which of two moving objects has greater speed.	3.5
8.EE.B.6 Use similar triangles to explain why the slope m is the same between any two distinct points on a non-vertical line in the coordinate plane; derive the equation $y = mx$ for a line through the origin and the equation $y = mx + b$ for a line intercepting the vertical axis at b.	3.5
Analyze and solve linear equations and pairs of simultaneous linear equations.	
8.EE.C.7 Solve linear equations in one variable.	3.1, 3.2, 5.2, 3.4
8.EE.C.7.a Give examples of linear equations in one variable with one solution, infinitely many solutions, or no solutions. Show which of these possibilities is the case by successively transforming the given equation into simpler forms, until an equivalent equation of the form $x = a$, $a = a$, or $a = b$ results (where a and b are different numbers).	3.2

Continued on next page

Common Core State Standards Alignment

COMMON CORE STATE STANDARD	FISHING ACTIVITY
The Number System	
Know that there are numbers that are not rational, and approximate them by rational numbers.	
8.NS.A.1 Know that numbers that are not rational are called irrational. Understand informally that every number has a decimal expansion; for rational numbers show that the decimal expansion repeats eventually, and convert a decimal expansion which repeats eventually into a rational number.	
8.NS.A.2 Use rational approximations of irrational numbers to compare the size of irrational numbers, locate them approximately on a number line diagram, and estimate the value of expressions (e.g., π^2). For example, by truncating the decimal expansion of $\sqrt{2}$, show that $\sqrt{2}$ is between 1 and 2, then between 1.4 and 1.5, and explain how to continue on to get better approximations.	1.3
Expressions and Equations	
Work with radicals and integer exponents.	
8.EE.A.1 Know and apply the properties of integer exponents to generate equivalent numerical expressions. For example, $3^2 \times 3^{-5} = 3^{-3} = 1/3^3 = 1/27$.	1.2, 1.3
8.EE.A.2 Use square root and cube root symbols to represent solutions to equations of the form $x^2 = p$ and $x^3 = p$, where p is a positive rational number. Evaluate square roots of small perfect squares and cube roots of small perfect cubes. Know that $\sqrt{2}$ is irrational.	5.4, 5.5, 5.7, 1.3

Continued on next page

checked their work. Conversation could be also initiated via the Forum as well.

Another approach you might consider is sampling one or two groups' work and emphasize a particular aspect of their strategy, in order to have specific questions answered (either as a whole class or in each group) that pertain to their specific approaches, procedures, methods, or errors where produced. For example, if a particular group arrived at a solution graphically, how could they do it algebraically. If errors were made although the reasoning was correct, how could the errors have been avoided?

Wrap-up (5-10 minutes)

As we transition from the final sections, a class discussion might conclude the NKH Activity Set by comparing advantages and disadvantages of the approaches in the activity. Such discussion may center around shared difficulties or possible shortcuts that students may have developed together or independently. It is important to recognize students' feelings and attitudes both during and after these activities.

Please note that the timing for these sections and activities range from 10 to 20 minutes, but the timing will vary from classroom to classroom depending on the nature of your needs within your particular instructional setting.

Additional Notes

Please feel free to spend as much time on these activities for your classroom. Finally, the ▸ icon in the text indicates where students can use the online learning platform.

Please send us any comments, issues, technical or otherwise, you might have with the content, the format or the approach.

References

Hattie, J., & Timperley, H. (2007). The power of feedback. Review of educational research, 77(1), 81-112.

Mathematics Assessment Resource Service; University of Nottingham & UC Berkeley; http://map.mathshell.org/index.php.

Remember to provide prompts to your haumāna continuously, for example:

- Recall what we were working on previously. What was the task about?
- I have had a look at your work. I would like you to think about the questions I wrote.
- On your own, carefully read through the questions I have written. I would like you to use questions to help you think about ways of improving your work.
- Use your mini-whiteboards to make a note of anything you will think of that may improve your work.

It is imperative to have your students review their own work before working collaboratively in groups. When working independently, you may tailor to individual students' suggestions that might help them clarify what they are thinking.

Collaborative Small-Group Work

After each student has been allowed time to work on their own problem, they can address the teacher's guiding prompts as a group to get a better understanding of the problem and reach a solution.

During this time, you may provide a few additional guiding prompts for your student groups:

- You each have your own individual solution to the task and have been thinking about how you might improve it using the questions I have posed. Now, I want you to share your work with your partner(s). Take turns to explain how you did the task and how you think it could be improved.
- If explanations are unclear, ask questions until everyone in the group understands the individual solutions.

This leads to the next few minutes where students are jointly working on a common solution. If you are using the NKH website, your joint solution can be posted onto the Forum. Notice that both of these formats will allow for sharing, discussing and analyzing the different groups' approaches towards a solution. Of course, the use of your own professional judgment on any other format achieving this goal will suffice.

During the section, you should take note of the different approaches between groups, the change of directions and dialogue between groups, etc. This effort will help you guide the whole-class discussion wrap-up.

Class management might/should be different from usual. This could be challenging. Remember that groups are like teams, and they will present and defend their team work. This might help keep students on task while having fun. Who said that was impossible?

Additionally, you'll support and foster problem-solving skills by asking questions that help your haumāna clarify their thinking, promote further progress, and encourage students to develop self-regulation as well as error detection skills.

Sharing, discussing, analyzing different approaches (10-20 minutes)

A whole-class discussion could follow the previous section. If posters were created, voluntary or randomly designated groups could share their strategies that were developed towards a joint solution. It may be important to ask how the students' group solution differed from their individual solutions. If your students do not explicitly state their conclusions, you might ask how they

Common Issues	Feedback Examples
Student has difficulty getting started	• What do you know? • What do you need to find out?
Omits some given information when solving the problem	Write the given information in your own words.
Overlooks or misinterprets some contraints	• Can you organize ... in a systematic way? • What would make sense to try? Why? • How can you organize your work?
Makes incorrect assumptions	Will it always be the way you describe?
Work is poorly presented	• Could someone unfamiliar with the task easily understand your work? • Have you explained how you arrived at your answer?
Provide little or no justification	How could you convince me that ... ?
Completes the task early	• How can you be sure that ... ? • What would happen if ... ? • Is there a way of describing all solutions?
Has technical difficulties	• Double check your work • Does it make sense? • Can you spot any mistakes?

A great deal of pedagogical creativity could take place with the integration of these activities and feedback techniques. A highly recommended research reference on giving feedback to students is Hattie and Timperley (2007).

During the Lesson

When you are delivering the **NKH SBAC aligned problems**, please make sure that you have allotted enough time for the students to work on the problem.

That being said, reviewing a student's first attempt at a problem should take a few minutes. This review is done individually, then general help and guidance can be provided to the entire class, as a collective whole, by either projecting on a screen or writing on the board.

Before the Lesson

Pre-Lesson Attempt *Optional*

After completing a Unit, please remember to give each student a copy of the appropriate **NKH SBAC aligned problem** or give them the password needed to access the activities on the NKH website. Some teachers may choose to give out these problems ahead of a lesson and see how much their students know. This effort will provide you with an opportunity to assess students' work in advance to find out the kinds of difficulties students have with the problems. With this knowledge, you should then be able to target your interventions and strategies to effectively help students in the subsequent problems.

Examples of instructional prompts for individual student who are attempting any given NKH problem include the following:

- Read the questions and try to answer them as carefully as you can.
- Show all your work, so that I can understand your reasoning.
- In addition to trying to solve the problem, I want you to check if you can present your work in a clear and organized way.

Students should work individually and without your assistance. Note, you may have to rearrange your students' seating arrangements. However, this instructional strategy will allow you to have a more accurate and informative picture of your students' understanding and present levels of performance.

Assessing students' responses and giving feedback

After collecting students' attempts at the *Let's Go Fishing* **NKH SBAC aligned problems**, please take the time to create a few notes for yourself (and as potential feedback) about what the students' sample work reveals about their *current levels* of understanding, as well as their individual and different approaches to problem-solving.

Scoring is not recommended during this phase.

It is also important to note that as a kumu, your feedback should summarize students' difficulties as a series of questions. For example, you can do this by:

- Writing one or two questions on each student's work;
- Giving each student a printed version of your list of questions and highlight the questions that are more relevant to each individual student; or
- Selecting a few questions that will be of help to the majority of students and sharing them collectively with the whole class (either projected or written on the board) when returning student NKH problem set first attempts.

When providing feedback to your haumāna, please refer to your own professional judgment and the respective needs of your individual instructional setting. That being said, certain common issues do arise across different classrooms and we recommend the following instructional prompts:

3. Next, students can work in small groups to combine their thinking—four students per group is often desired, as students could also work in pairs, within the small group. They work together to produce a collaborative solution to the given question(s).
4. In the same small groups, students evaluate and comment on each others' responses, identifying the strengths and weaknesses in these responses and comparing them with their own work.
5. In a whole-class discussion, students compare and evaluate the strategies they have seen and used.
6. In a follow-up lesson, students reflect for 10 minutes on their work and what they have learned.

Materials Required

- **NKH SBAC aligned Let's Go Fishing workbook** or a **computer** with access to the STEMD2 website (www.stemd2.com)
- A **Mini-Whiteboard** with **Marker** and **Eraser** for a quick way to visibly check individual understanding (this instructional strategy also enhances attention and participation)
- **Large sheets of paper** for small groups to create posters, if desired
- **Calculators and Graph Paper** for certain activities
- **Projector and Screen** to share students sample response

Time Needed

- **Pre-Lesson Attempt:** 5-10 minutes
- **Follow-up and Reflection:** 5-10 minutes

Lesson Planning Structure

Introduction

Aloha kākou, e komo mai. Welcome to our **Neʻepapa Ka Hana (NKH)** guide for kumu! In *Let's Go Fishing*, we have developed a series of questions that model SBAC test questions to prepare your students for the SBAC (Smarter Balanced Assessment Consortium) testing at the end of the school year.

Please utilize your professional judgment to determine when and where to integrate these activities directly into your lessons or before SBAC testing. Following this Introduction, we have also provided an outline of this book's CCSS (Common Core state Standards) alignment to help guide you and the integration of these questions into your classroom.

Perhaps, these activities may serve as a culminating activity or Summative Assessment at the end of your lessons. Alternatively, it may also be appropriate to incorporate the NKH SBAC aligned questions as a Formative Assessment. Either way, the NKH SBAC aligned questions can serve as a supplementary bridge to your mathematics curricula to enhance **Inclusive Classroom Pedagogy**, and **Problem Based Learning**.

The problems presented in the NKH SBAC aligned activities incorporate real-world problem solving tasks that reflect Hawaiian society, geography, and culture. We developed these activities with a goal for both kumu and haumāna to collaborate, communicate together, and think critically and creatively—and thus, deepening understanding, application, and appreciation for mathematical thinking.

Formative Steps in the NKH Activity Set

In *Let's Go Fishing*, each question can generally be used independently. The activities can serve as a cumulative activity that ideally should be considered after the various components of a given unit have been covered in class. We recommend the following steps in completing the activities in this book:

1. Students may work individually on the questions.
2. The students' answers and process can be viewed by the teacher to gauge students understanding.

Preface

About the STEMD² Book Series

The STEMD² Book Series for eighth-grade mathematics was developed as part of a technology-enabled pedagogical approach (Neʻepapa Ka Hana model) for teaching mathematics in Hawaiʻi middle schools. This book series seeks to provide Hawaiʻi middle school teachers resources and training to incorporate problem-based learning, social learning and inclusive pedagogy through a culturally relevant mathematics curriculum. The series currently includes:

> **Let's Build a Canoe** – Student Activities and Teacher's Guide (Common Core aligned)
> **Let's Play the ʻUkulele** – Student Activities and Teacher's Guide (Common Core aligned)
> **Let's Go Fishing** – Student Activities and Teacher's Guide (SBAC aligned)
> **Let's Make Da Kine / E Hana Kākou** – Student Mini Projects and Teacher's Guide in English and ʻŌlelo Hawaiʻi (Skill development)

The printed and online resources produced by NKH through STEMD² are fully aligned with the Common Core Standards for Mathematical Practice and Content and the Smarter Balanced Assessments for mathematics. Based on the on GO Math!® curriculum structure, the NKH STEMD² book series and social learning platform (community.stemd2.com) is flexible for teachers to implement partially or fully in their classrooms, as a tool to encourage students' interest and achievement in STEM subjects.

Funded by a three-year grant from the Department of Education's Native Hawaiian Education Act Program, Neʻepapa Ka Hana is a project of the STEMD² Research & Development Group in the Center on Disability Studies at the University of Hawaiʻi at Mānoa. More information about STEMD² is available online at www.stemd2.com.

Activity 3.4 42
Activity 3.5 43

Unit 4: Transformational Geometry 45
Activity 4.1 46
Activity 4.2 47
Activity 4.3 48
Activity 4.4 49
Activity 4.5 51

Unit 5: Measurement Geometry 53
Activity 5.1 54
Activity 5.2 55
Activity 5.3 56
Activity 5.4 57
Activity 5.5 58
Activity 5.6 59
Activity 5.7 60
Activity 5.8 61
Activity 5.9 62

Unit 6: Statistics 63
Activity 6.1 64
Activity 6.2 66
Activity 6.3 68

Contents

Preface 7

Lesson Planning Structure 9
Introduction 9
Materials Required 10
Time Needed 10
Before the Lesson 11
During the Lesson 12
Wrap-up 14
Additional Notes 14
References 14

Common Core State Standards Alignment 15

Unit 1: The Number System 21
Activity 1.1 22
Activity 1.2 23
Activity 1.3 24
Activity 1.4 25

Unit 2: Proportional and Nonproportional Relationships and Functions 27
Activity 2.1 28
Activity 2.2 29
Activity 2.3 31
Activity 2.4 32
Activity 2.5 33
Activity 2.6 34
Activity 2.7 35

Unit 3: Solving Equations and Systems of Equations 37
Activity 3.1 38
Activity 3.2 39
Activity 3.3 41

Neʻepapa Ka Hana Eighth-Grade Mathematics Resources
Let's Go Fishing
Teacher's Guide

Project Directors	Kaveh Abhari Kelly Roberts
Content Developers	Justin Toyofuku Robert G. Young
Publication Designers	Katie Gao Robert G. Young MyLan Tran
Content Contributors	Remy Pages Kimble McCann

Acknowledgments

We would like to thank Ruth Silberstein, Cody Kikuta, Katy Parsons, Crystal Yoo, and Kelli Ching for advising on eighth-grade mathematics education.

Suggested Citation

Let's Go Fishing (2020). Neʻepapa Ka Hana Eighth-grade Mathematics Resources. STEMD² Book Series. STEMD² Research and Development Group. Center on Disability Studies, University of Hawaiʻi at Mānoa. Honolulu, Hawaiʻi.

Neʻepapa Ka Hana | STEMD2 R&D Group | stemd2.com

Made in the USA
Las Vegas, NV
27 September 2024

95803356R00043